Series / Number 90-025

Environment, Energy, and Black Americans

JOHN M. OSTHEIMER
LEONARD G. RITT

Northern Arizona University

SAGE PUBLICATIONS / Beverly Hills / London

Printed in the United States of America

For information address:

SAGE PUBLICATIONS, INC.
275 South Beverly Drive
Beverly Hills, California 90212

SAGE PUBLICATIONS LTD
St George's House / 44 Hatton Garden
London EC1N 8ER

International Standard Book Number 0-8039-0545-9

Library of Congress Catalog Card No. 75-9049

FIRST PRINTING

When citing a Research Paper, please use the proper form. Remember to cite the series title and include the paper number. One of the two following formats can be adapted (depending on the style manual used):

(1) SWAN, J. A. (1975) "Returning to Our Roots: Organic Education and Social Transformation," Sage Research Papers in the Social Sciences (Human Ecology subseries; No. 90-022). Beverly Hills and London: Sage Publications.

OR

(2) Swan, James A. 1975. *Returning to Our Roots: Organic Education and Social Transformation.* Sage Research Papers in the Social Sciences, vol. 3, series no. 90-002 (Human Ecology subseries). Beverly Hills and London: Sage Publications.

Contents

ENVIRONMENT, ENERGY, AND BLACK AMERICANS

JOHN M. OSTHEIMER
LEONARD G. RITT

Northern Arizona University

OVERVIEW

When the environmental movement reached its peak, culminating in the impressive legislative achievements of the 1969-1973 period, there was great hope that these issues could bring Americans together for a common purpose. Editorialists were quite optimistic about the chances for an alliance of old and young, rich and poor, black and white, liberal and conservative, based on the premise that environmental problems threatened all Americans with their insidious, if undramatic, growth (Tognacci et al., 1972: 74). It is obvious in retrospect, however, that not all Americans reacted in the same fashion.

Of extraordinary interest has been the reaction of many black people, particularly their leaders. The agenda of the 1972 National Black Political Convention in Gary, Indiana, included only one brief mention of air and water pollution. Convention participants seemed fatalistic about the prospects for environmental improvement, and preferred to concentrate on what seemed to them, more fundamental problems, arguing that such matters as pollution "will not be significantly affected by new faces in old places in Washington, D.C." (*Black World,* 1972). Although this convention took place during a time when considerable environmental activity was culminating at the international level with the 1972 Stockholm conference, the black community in this country seemed oblivious to these events (see also Strickland, 1972, and Walters, 1972).

AUTHORS' NOTE: The authors are grateful for the support provided for this project by Northern Arizona University and by Transition Foundation; we also wish to thank Elizabeth Collier for assisting with the research.

At the same time, while the number of blacks in city halls and in Congress has also been increasing, none of these officials has been among the country's noted environmentalists. In a 1970 survey, not one of twenty black mayors mentioned environmental problems as being of major importance (*Ebony,* 1970; see also 1973 and 1974). Black congressmen have not been among the sponsors of environmental legislation, nor have they been vocal or politically visible on these issues. Black magazines have virtually ignored ecological debate. For example, Alex Poinsett's column in *Ebony*, which offers a sampling of the political issues that concern American blacks, has not dealt at all with environmental concerns (Poinsett, 1973a).

Thus, according to Walter Rosenbaum (1973: 74), "ecology remains a peripheral concern to most disadvantaged Americans because their more urgent needs are adequate employment, decent diets, better housing and education, health care and other essentials." Whitney Young, former head of the National Urban League, seems to agree:

> The war on pollution is one that should be waged after the war on poverty is won. Common sense calls for reasonable national priorities and not for inventing new causes whose main appeal seems to be in their potential for copping out and ignoring the most dangerous and most pressing of our problems [Rosenbaum, 1974: 74].

Other black leaders have echoed this sentiment. The mayor of Gary, Richard Hatcher, was quoted stating: "The nation's concern with the environment has done what George Wallace was unable to do: distract the nation from the human problems of Black and Brown Americans" (*Time Magazine,* 1970). A study of the attitudes of 28 black students attending Michigan State University supports the views of Hatcher and Young. One student offered: "We did not destroy the environment—the white man did. Let him straighten up what he destroyed." Another said: "Some of the things ecologists advocated, black people have been doing for years—like recycling disposable materials" (Kreger, 1973: 31).

Some black leaders and intellectuals have gone even further in their criticisms. Comedian Dick Gregory, a popular figure among many younger black intellectuals (Salaam 1974), subscribes to a sort of devil theory to explain the emphasis placed by white Americans on environment. He labelled birth control programs "obvious tactics of genocide. . . . There is ample evidence that government programs . . . emphasize birth control and abortion availability, both measures obviously designed to limit the black population" Gregory: 1971.[1] On a more specific level, Dr. Charles

Greenlee, a black physician, defended the discontinuance of Pittsburg's Planned Parenthood clinics. He accused social workers of forcing poor black women to use the clinics by threatening to reduce or eliminate their welfare payments, arguing that "the free clinics constitute 'genocide,' a conscious conspiracy by whites to effect a kind of Hitlerian solution to the 'black problem' in the U.S." (Hallow, 1969; see also Walters, 1974, and Watts and Free, 1973: 154).

The controversy over compulsory sterilization has added to the conclusion among some black leaders that white America has racism in mind when it talks about population control. "One doesn't need to know much about involuntary sterilization . . . to detect that racism is involved and beyond mere racism, some unspeakable wish to control black population growth by force" (Slater, 1973; see also Brody, 1974). Many blacks react to this issue as they have before to the specter of frontal lobotomies performed on blacks convicted of violent crimes (Mason, 1973). Thus, because environmentalists stress population as an important factor in the deterioration of the world's natural environment, blacks have drawn some unpleasant deductions about environmentalists.

Of course not all blacks agree with the extreme conclusions of Greenlee and Gregory. John Hudgins has argued in favor of black efforts to control their own population. His view, however, was not based on "environmental" considerations, at least as white America defines them. Blacks must struggle, he says, for political power, and must be ready to defend their communities from white oppression. Their defense may need to be violent. Under such conditions, "when one looks closely at the situation in our community . . . can we really deal with more children? My answer has to be a qualified, no" (Hudgins, 1972).

There is, however, some evidence that black people—while they may find the accusing rhetoric of Gregory and other "genocide" alarmists amusing—are not planning to heed Gregory's advice. Between 1965 and 1973, census data show that the birth expectations of young black women, aged 18-34, have fallen more sharply than among white women (see Table 1). The number of births expected is still, for the most part, higher than for whites, but among the youngest group, 18-24—a group one would expect to be the most influenced by Gregory's sensational arguments—black women were expecting the *same* number of births as white women by 1973.

Of course, these data represent "realistic" expectations, rather than evaluations of what an "ideal" number of children would be. Gallup had asked just such a question, and the results give a very different impression. Among "non-whites," 37 percent answered that four or more children would be ideal, as opposed to only 15 percent of whites (Gallup,

TABLE 1

AVERAGE NUMBER OF BIRTHS TO DATE, AND TOTAL BIRTHS EXPECTED, BY WIVES
18 TO 39 YEARS OLD (US, BUREAU OF THE CENSUS, 1974) 1965-1973, BY RACE

	18 to 24 years		25 to 29 years		30 to 34 years		35 to 39 years	
Year and Race	Births to date	Total births expected	Births to date	Total births expected	Births to date	Total births expected	Births to date	Total births expected
WHITE								
1973	0.8	2.3	1.7	2.4	2.6	2.8	3.1	3.2
1972	0.9	2.2	1.8	2.4	2.7	2.8	3.1	3.2
1971	0.9	2.4	1.9	2.6	2.8	2.9	3.1	3.2
1970	1.0	2.6	2.1	2.7	2.8	3.0	3.2	3.2
1967	1.1	2.9	2.3	3.0	3.0	3.2	3.1	3.2
1965	1.3	3.1	2.4	3.3	3.0	3.5	3.1	3.3

TABLE 1 (Continued)

Year and Race	18 to 24 years		25 to 29 years		30 to 34 years		35 to 39 years	
	Births to date	Total births expected	Births to date	Total births expected	Births to date	Total births expected	Births to date	Total births expected
Blacks								
1973	1.4	2.3	2.2	2.8	3.1	3.3	3.9	3.9
1972	1.3	2.4	2.1	2.8	3.5	3.7	3.9	4.0
1971	1.4	2.6	2.5	3.1	3.5	3.7	4.1	4.2
1970	1.7	2.9	2.6	3.2	3.7	3.8	3.9	4.1
1967	1.8	2.8	3.0	3.4	3.9	4.3	4.2	4.2
1965	2.1	3.4	3.4	4.0	3.8	4.4	3.8	4.1
Change from 1965-1973								
Whites	-.5	-.8	-.7	-.9	-.4	-.7	-0	-.1
Blacks	-.7	-1.1	-1.2	-1.2	-.7	-1.1	.1	-.2

1974b: 28). Time-series data were not available for blacks, but the white public's decline from roughly 40 percent answering "four or more children would be ideal" in 1968, to only 15 percent six years later, certainly reflects a rapidly growing preference for small families. Unfortunately, we do not know whether black attitudes toward ideal family size have changed similarly, but it seems unlikely that they could have changed as rapidly as did attitudes among whites.

From these admittedly confused results, we conclude tentatively that there seems to be a gulf between the approach taken by many whites and blacks on environmental issues. Could the words "environment" and "ecology" connote different things for black people? This is part of the answer. Environmental problems *are* different for blacks: they are reflections of the socioeconomic deprivation caused by inequalities in housing, education, health care, and other basic amenities. The idea of "environment" exists in the contemporary thinking of black Americans within a context of social, economic, and sometimes physical oppression (Baker, 1971). To Nathan Hare (1970), editor of *The Black Scholar,* "the concept of ecology is potentially of momentous relevance to the ultimate liberation of black people." But Hare admits that, as things stand now, blacks see the ecology movement as a problem for white society, partly because white environmentalists have ignored the issues that *are* of real concern to black people.

One of the most eloquent treatments of the gap between the interests of blacks and those of the environmentalists is found in a 1973 symposium on the topic of "growth." Willard Johnson, black political scientist at the Massachusetts Institute of Technology, strongly chided the environmental movement for its lack of attention to the problems of the poor (Johnson, 1973). Johnson's reference to Anthony Crosland's criticism of environmentalists, who, though well meaning, are "affluent; and fundamentally, though of course not consciously . . . want to kick the ladder down behind them" set the stage for his analysis (1973: 165). Johnson was clearly unimpressed by either the economics or population oriented "zero-growth" arguments. In his judgment the view of population level as an environmental problem was in itself incorrect and irrelevant.

> There is plenty of room for population growth in the United States . . . the real population issues to the poor are the incompatibility of their values and desires with those of the rich, and their lack of power to protect and satisfy their own values [1973: 166-167].

As far as economic growth is concerned, Johnson agreed with the environmentalists up to a point: economic growth was not accurately justified on the grounds that it finances improving status for the poor or blacks. He presented data that belied any such interpretation of the benefits of growth (1973: 169, 171-178). But then Johnson's argument parted company with the environmentalist "no-growth" advocates because he could find no assurance that limited growth would not affect black people with a severity the environmentalists chose to ignore. Indeed, to the extent that environmentalists desire a stop to growth because of ecological, rather than aesthetic or nostalgic reasons, the concept does present a clear threat to blacks:

> The industries in which blacks have a foothold and a potential for economic improvement are precisely those where we find the greatest ecological hazards. Blacks are particularly entrenched in the auto industry and many of the industries and services peripheral to it. Blacks constitute 23 percent of nonfarm, nonconstruction laborers. Black workers form substantial contingents in industries which deal with chemicals, fabricated metals, primary metals, and nonelectrical machinery and transportation equipment. Industries like these are closely identified with the problems of poisoned lakes and streams, and with overuse of material resources, especially of nonrenewable resources such as petroleum, natural gas, and other fueles that will become increasingly important as energy sources in the future. . . . Those whose predominant objectives are ecological may threaten economic progress for blacks as well as for great numbers of nonblack and nonminority poor [1973: 170-171].

In short, no-growth advocates, "for all their humaneness, ignore the concerns of the poor" (1973: 179).

The best counterarguments environmentalists can offer (few do so directly) are restatements of the severe effects that urban density, ordinarily associated with growth, has on people. The recent arguments over contaminated urban water supplies, and the continuing threat that urban air quality standards will be lowered under the pressure of energy shortages, provide evidence that growth can hurt blacks, along with everyone else.

Nonetheless, the prevailing attitude among black elite is that environmentalists do not have the black community's interests at heart.

Nathan Hare has effectively summarized the differences between blacks and the environmental movement during the mid-1970s. In contrasting black and white ecological perspectives, Hare noted three points of departure:

(1) Black and white environments not only differ in degree but in nature as well;
(2) The causes and solutions to the ecology problems are fundamentally different in the suburbs and ghetto;
(3) The solutions set forth . . . are reformist and evasive of the social and political revolution which black environmental correction demands.

Up to this point we have concentrated attention on statements by black political and intellectual elites indicating what blacks do (or should) think about environmental issues. What, in actuality, do black masses believe about environmental problems? Furthermore, how do congressmen who represent predominantly black districts respond to these issues? The first question will be explored through recent survey data; the second by analyzing congressional voting behavior in the 92nd and 93rd Congresses.

ENVIRONMENTAL ATTITUDES OF THE BLACK MASSES

During the peak of the environmental movement, the typical evaluation of black mass attitudes toward environmental problems was that "the poor, the black, and those with only a grade school education have been least likely to care about improving environmental quality" (Trop and Roos, 1971: 53; see also Tognacci et al., 1972). Table 2 verifies such an assessment. Some 21 percentage points separate blacks from whites, who were far more willing to commit more government funds to environmental cleanup (Harris, 1969: 25).[2] What is equally revealing, however, is that over twice as many blacks as whites answered "not sure" on this particular question—a good indication that the issue lacked salience for this group of Americans.

Ordinarily, differences in socioeconomic status are cited to explain such racial discrepancies, and this has certainly been the case in analyses of environmental attitude tata. As Clarence Davies (1970: 80) put it, "for both air and water pollution, concern with the problem increase as one moves higher up the socioeconomic scale. The higher a person's income, the more education he has had, and the higher the status of his occupation, the more likely he is to view air and water pollution as serious problems" (see also Trop and Roos, 1971: 85 and Loveridge, 1974: 9). Thus, "mainstream" logic would dictate that as black people become more aware of (that is, socialized through education and the media to the concept of) ecological interdependence, the astounding 43 percent who are "not sure" would naturally choose to commit funds to the cleanup, and would re-

TABLE 2

ATTITUDES TOWARD NATIONAL SPENDING ON ENVIRONMENT, BY RACE

"Thinking about air and water pollution, improvement of land and water, forest, fish and wildlife, recreation and park areas----- do you think programs for improvement of the natural environment now receive too little attention and financial support from the government, or just the right amount?" (Louis Harris and Associates, 1969:5).

	Receive Too Little	Receive Too Much	Receive Right Amount	Not Sure	
White	54%	5%	22%	19%	100%
Black	33%	6%	18%	43%	100%

N=1485

spond to family planning. Indeed, the authors did find some evidence that, with greater education, the preference for smaller families increases among blacks, as among whites (see Table 3). But, when we turned to the question of willingness to pay for pollution cleanup, the theory that increasing socioeconomic status among blacks will result in greater environmentalism among the more informed segment of the black population was *not* confirmed by our analysis. The reverse appeared to be the case. Table 4, based on National Opinion Research Center data, shows that college educated blacks are *less* willing to approve spending for environmental cleanup than are their white counterparts or their less educated brethren. As expected, the inclination among more educated blacks is to support more public spending on health programs, urban problems, and education, but this trend is not manifested in response to environmental issues. This conclusion draws strong support from data collected by Watts and Free (1973), as presented in Tables 5 and 6. Hence, it would appear that at least for blacks, increased environmental awareness is not simply a function of increased levels of information.

One might argue—given the small number of college-educated blacks in such samples that these findings must be treated with caution. There is, however, more corroborative evidence in Table 7, which suggest that knowledge of environmental problems heightens black ambivalence. It may be that the information sources of more educated blacks are those very spokesmen who denounce environmentalism as a "cop out." On the other hand, it is possible that the issue salience of environmental problems, *at least as white American culture has defined them,* is low. The more the average black American comes to know about such concerns, the more likely he is to deemphasize environmental cleanup in favor of stressing policy areas of greater perceived importance to the black community. The findings of Swan (1972: 70) strongly support this interpretation: Detroit high school students were highly interested in issues related to employment and the functions of the police, and were eager to participate in decision-making activity on those issues. However, "faced with the reality of more immediate problems like employment and housing, air pollution is likely to receive a lower priority. . . . It is little wonder that so few Blacks were participants in Earth Day."

BLACK CONGRESSMEN AND THE ENVIRONMENT

Interpreted in this manner, black public opinion data seem to confirm the impression given by Hatcher, Young, and other black leaders, that environmentalism has not been as popular a topic within America's black

TABLE 3

Question: "What do you think is the ideal number of children for a family to have?" (National Opinion Research Center, 1972:27) By race and education.

Educational Level	Less than High School				High School				More than High School			
Ideal Number of Children	One or two	three or four	five or more	Totals	One or two	three or four	five or more	Totals	One or two	three or four	five or more	Totals
White	176	232	37	445	312	265	29	606	112	51	3	166
	40%	52%	8%	100	51%	44%	5%	100	67%	31%	2%	100
Black	31	42	32	105	37	46	6	89	4	4	0	8
	30%	40%	30%	100	42%	52%	6%	100	50%	50%	0	100

TABLE 4

Black and White Attitudes toward Spending on Major National Problems, by race and education.

% who answer we are spending "too little" to solve the particular problem

	Health Programs	Cities	Education	Defense	Foreign Aid	Environment	
Whites							
No College	59	50	44	13	4	61	N=824
College	68	57	55	8	3	73	N=408
Blacks							
No College	72	66	71	22	11	68	N=128
College	78	76	86	9	3	65	N= 37
							N-1397

Source: National Opinion Research Center, (1973).

TABLE 5

"Worries and Concerns" - Domestic Issues, by race

Q. "Now I'd like to find out how worried or concerned you are about a number of problems I am going to mention: a great deal, a fair amount, not very much, or not at all. If you aren't really concerned about some of these matters, don't hesitate to say so." (WATTS and FREE, 1973:288).

	White	Black
Crime	89	90
Drugs & Addicts	90	88
Violence	90	91
Housing	66	87
Consumer Protection	82	82
Education	76	86
Poverty	75	90
Inflation	90	92
Unemployment	76	90
Business Conditions	70	80
Medical Care	82	87
Water Pollution	84	80
Air Pollution	83	80
Solid Wastes	80	81
Urban Renewal	71	86
Black Americans	62	90
Mass Transportation	55	70
Senior Citizens	77	84
Urban Problems	70	83

Source: WATTS and FREE, 1973:315. (Data indicate an "index of high concern." See *Ibid.* 356)

TABLE 6

Government Spending - Domestic Issues, by race

Q. "Now I'm going to read off the names of some programs the federal government in Washington is helping to finance. As I mention each program, please tell me whether you feel the amount of tax money now being spent for each should be increased, kept at the present level, reduced, or ended altogether." (WATTS and FREE, 1973:294)

	White	Black
Public Housing	60	83
Housing in General	61	88
Educating Low-Income Children	78	91
Higher Education	70	89
Highways	63	65
Urban Renewal	66	88
Water Pollution	81	76
Air Pollution	81	75
Sr. Citizens	86	95
Medicaid	72	84
Black Americans	53	95
Mass Transportation	65	73
Welfare	49	81
Medical Care	78	91
Crime	88	88
Drugs & Addicts	86	88
Jobs for Unemployed	65	85
Public Parks	65	78
Urban Problems	66	85
Space Program	26	14

Source: WATTS and FREE, 1973:332. (Data indicate an "index of willingness to spend." See *Ibid*. 356-7).

TABLE 7

Willingness to see Federal Monies Spent on Natural Resources (environmental improvement, forests, recreation areas, wildlife, pollution control, etc.) by race.

"Considering priorities, would you like to see more or less of the federal money go into each of these purposes?" (900 respondents were shown a list of 10 federal budget categories, with percent of the total spent on each category. Approx. 500 were shown the list, but <u>without</u> the percentage figures. "Natural Resources" received 1% of the allocations that year.)

	Spend more on Natural Resources	Spend less on Natural Resources	Not Sure	
White	63%	13%	24%	100%
Black	52%	8%	40%	100%
	With current budget allocation figures (N=900)			
White	71%	9%	20%	100%
Black	35%	12%	53%	100%
	Without current budget allocation figures (N=500)			

Source: Harris, L. and Associates, Inc. (1969:22-24).

community as it has been for whites. But do the expressions of distrust for the issue by black leaders, coupled with the black public's apathy toward it, translate into a hostile legislative performance by the Congressional Black Caucus?[3] Apparently not: Table 8 demonstrates that the voting records of black congressmen, as evaluated by the League of Conservation Voters (1961-1973) are impressively "environmentalist." Of 44 "rating years" (number of blacks congressmen *times* years scored by League of Conservation Voters) compiled between 1970-1973, only Congressman Nix (D-Pa) shows a percent-correct score of less than 50 percent, and for only one of the years concerned (1971). Perhaps black congressmen do not stand out as spokesmen for environmental initiatives, and make few speeches on the topic, but their voting records are, nevertheless, remarkably in accord with the League of Conservation Voters' positions on major environmental legislation.

How can this obvious discrepancy between black attitudes outside Congress and black congressional voting records be explained? More detailed analysis of the LCV rating system helps to unravel a portion of this apparent irony. Congressional rating systems, including even the "special interest group" ratings, are rather selective collections from among the total votes considered by Congress each session. Any particular bill may be promoted by a variety of interest groups, perhaps in active alliance, but for very different reasons. The LCV rating is no exception: LCV often overlaps with other interest groups in its choice of votes. For example, black congressmen seem to have been unanimously "pro-environment" in voting against the Dennis and Findlay amendments to the Occupational Safety and Health Act (June 15, 1972), which would have: (1) reduced Labor Department enforcement personnel, and (2) temporarily exempted smaller firms from the Act's provisions. But were black congressmen voting "pro-environment" or "pro-labor?" The authors have elsewhere examined congressional voting on environmental issues, and found that high correlations exist among a group of congressional rating systems one might call "liberal" (Ritt and Ostheimer, 1974). LCV's inclusion of "workplace environment" votes strengthened the degree of overlap with labor rating scores. One, therefore, needs to examine the separate votes that comprise LCV scores in order to determine how often black congressmen vote "pro-environment" for reasons that have to do with environmental concerns *in their own right.* For example, in 1972, absenteeism and anti-environmental voting increased markedly among black congressmen for the Dickey-Lincoln Dam vote, a "wilderness-conservation issue" (see vote 9, Appendix A.[4] Table 9 gives an index of agreement between black congressmen and environmentalists' positions for each issue during the 92nd Congress. This index was constructed by subtracting the number of black

TABLE 8

Black Congressmen's Voting Scores on Environmental Issues: 1970-1973
Percent Voting "Environmentally Correctly"

		1970	1971	1972	1973
Hawkins	(D-CA)	65	63	86	82
Dellums	(D-CA)	--	93	100	100
Burke	(D-CA)	--	--	--	93
Young	(D-GA)	--	--	--	100
Metcalfe	(D-IL)	--	52	70	61
Collins	(D-IL)	--	--	--	73
Mitchell	(D-MD)	--	86	98	95
Conyers	(D-MI)	90	60	94	85
Diggs	(D-MI)	80	70	90	79
Clay	(D-MO)	100	54	88	90
Chisholm	(D-NY)	90	83	80	96
Rangel	(D-NY)	--	89	94	98
Stokes	(D-OH)	100	83	94	90
Nix	(D-PA)	53	47	85	61
Jordan	(D-TX)	--	--	--	79
Average LCV Score for the Black Delegation		83%	69%	88%	85%

Source: League of Conservation Voters (1961-1970, 1971, 1972, and 1973).

TABLE 9

Environmental Voting Records of Black Congressmen
92nd Congress (League of Conservation Voters, 1971, 1972)

	Hawkins	Dellums	Metcalfe	Conyers	Diggs	Clay	Chisholm	Rangel	Stokes	Nix	Index of Environmental Voting[a]
1971											
1. Family Planning[b]	E	a	a	E	E	E	E	a	E	X	3.5
2. SST	X	E	E	E	E	E	E	E	E	X	6.0
3. SST	X	E	E	E	E	E	E	E	E	E	8.0
4. Mass Transit	a	E	E	a	a	E	E	E	E	a	4.0
5. Mass Transit	E	E	X	E	E	E	E	E	E	E	8.0
6. Channelization	E	E	E	E	E	a	E	a	a	X	3.5
7. Radioactive wastes	a	E	X	a	a	E	E	E	E	X	1.5
8. Amchitka blast	E	E	a	a	a	a	E	E	E	X	2.0
9. Dickey Lincoln dam	X	X	a	X	a	a	X	X	X	E	-6.5
10. Alaska wilderness	E	E	X	E	a	X	X	E	E	X	.5
11. Pesticides	E	E	a	a	a	a	E	E	E	E	4.0
12. Pesticides	E	E	a	a	a	a	E	E	E	E	4.0
13. Pesticides	E	E	E	a	a	a	E	E	E	E	5.5
14. Campaign practices	E	E	E	E	a	a	a	E	E	E	5.5
15. Campaign practices	X	E	X	X	a	X	E	E	X	X	-3.5

TABLE 9 (continued)

1972	Hawkins	Dellums	Metcalfe	Conyers	Diggs	Clay	Chisholm	Rangel	Stokes	Nix	Index of Environmental Voting[a]
1. Water pollution	a	E	X	a	E	E	a	E	E	E	3.5
2. Water pollution	a	E	X	E	a	a	a	E	a	E	.5
3. Water pollution	a	E	E	E	X	E	a	E	E	E	5.0
4. Water pollution	a	E	E	E	E	E	a	a	E	E	5.5
5. Water pollution	a	E	E	E	E	E	a	a	a	E	4.0
6. Recycle sewage	a	E	E	E	E	E	a	a	E	X	3.5
7. Gov't liability	E	E	a	a	E	a	a	E	E	E	4.0
8. Workplace envir.	E	E	a	E	E	E	E	E	E	E	8.5
9. Workplace envir.	E	E	a	E	E	E	E	E	E	E	8.5
10. Highway trust	E	E	E	E	E	a	a	E	E	E	7.0
11. Three Sisters bridge	E	E	a	a	E	a	a	E	E	X	2.0
12. Strip mine	a	E	E	E	E	a	a	E	a	E	4.0
13. Water pollution	E	E	E	E	E	a	E	E	E	E	8.5

a. This index was prepared by the authors, not by LCV
b. For more descriptions of each vote, See Appendix A.
E = "environmental" vote; X = "Anti-environment" vote; a = absent, not voting

congressmen who voted "unenvironmentally" on an issue from the number who voted in accord with the environmentalists. Absences were counted as one-half an "unenvironmental" vote. The results certainly show that black congressmen tend to support environmentalist positions less on wilderness-conservation issues.[5] It follows than that LCV ratings based more on these kinds of issues, as in 1971, will produce lower "environmental voting" records for black members of Congress. (Average 1971 LCV score for black congressmen = 69 percent.) The 1972 and 1973 higher average LCV scores are in part accountable to the replacement of wilderness issues by "workplace-environment" and "urban-problems" votes.

ENTER THE ENERGY CRISIS

During the 93rd Congress, events spawned a situation in which one would have expected the gap between black Americans and the environmental movement to widen. Energy shortages, coupled with the generally deteriorating state of the economy, awakened many blacks to the threat of job insecurity and their inability to keep up with the costs of living (Farrell, 1974).[6]

Being poor, concentrated in the inner cities, and therefore vulnerable as a group, blacks interpreted the energy situation as particularly dangerous to to them. From Chicago, Boston, Harlem and Baltimore, came reports of heat shortages during the winter of 1973-1974; these were accompanyed by demonstrations by angry crowds (Farrell, 1974). Furthermore, at the second National Black Political Convention in Little Rock, delegates discussed the adverse effects of energy shortages. This was in contrast to the 1972 Gary convention, which had ignored environmental concerns. Vernon Jordan (1974) summarized the fears of many blacks: "About the only certainty in the energy crisis situation is that black people will bear a disproportionate burden of the sacrifices and penalties" (see also Wilson, 1974, Seward, 1974, and *New York Times,* 1974).

Opinion data can be used again to illustrate how the reaction of black people differs from that of white Americans. Table 10 shows that blacks favor fewer controls on the consumption of energy; they are, for example, less sympathetic than whites to reduce highway speed legislation. Again, however, the low salience of this issue area is demonstrated by the fact that in September 1973, one-third of the blacks were not aware of the energy crisis. Black congressmen, nevertheless, showed a consistently strong environmental stand on all major issues (Table 11). This was true even though it was pointed out by some that the impact of many of the environmental initiatives during 1973 could result in price increases in the

TABLE 10

Contrasting Racial Attitudes toward Ways to Reduce Energy Consumption; by race

"Do you think controls on the use of energy, that is, gasoline, oil, and electricity--should be made more strict, less strict, or kept about as they are now?" (Gallup opinion index, 1974a:10).

	More Strict	Less Strict	About Same	No Opinion	
White	42%	5%	47%	6%	100%
Non-white	21	11	60	8	100

"It has been proposed that, as a way to conserve gas, speed limits on the nation's major highways be reduced by 10 miles per hour. Do you think this is a good idea or a poor idea?" (Gallup opinion index, 1973:37).

	Good Idea	Poor Idea	No Opinion	Not Aware of "Energy Crisis"	
White	54%	28%	4%	14%	100%
Non-white	33	26	7	34	100

TABLE 11

Voting Records of Black Congressmen; Selected Energy-environmental Issues from the 93rd Congress, 1973

		Dellums	Hawkins	Burke	Young	Metcalfe	Collins[a]	Mitchell	Conyers	Diggs	Clay	Chisholm	Rangel	Stokes	Nix	Jordan	Index of Environmental Voting
1.	Mass Transit[c]	E	E	E	E	E		E	E	E	E	E	E	E	E	E	14[b]
2.	UNEP	E	E	E	E	E		E	E	E	E	E	E	E	a	E	12.5[b]
3.	Population Growth	E	E	E	E	a	E	E	E	a	E	E	E	E	E	E	12
4.	Atomic Power Plants	E	E	E	E	E	X	E	E	a	E	a	E	E	E	X	8
5.	Project Plowshare	E	E	E	E	X	X	E	E	X	E	E	E	E	E	X	7
6.	Geothermal Energy	E	E	E	E	X	X	E	X	E	E	E	E	E	X	E	7
7.	Pesticides	E	E	E	E	E	E	E	a	E	E	a	E	a	E	E	10.5
8.	Toxic Substances	E	E	E	E	E	E	E	E	E	a	E	E	E	E	E	13.5
9.	Toxic Substances	E	E	E	E	E	E	E	E	E	E	E	E	E	X	E	13
10.	Alaska Pipeline	E	X	E	E	E	E	E	E	E	E	E	E	E	X	E	11
11.	Alaska Pipeline	E	X	E	E	X	E	E	E	E	E	E	E	E	E	E	11
12.	Alaska Pipeline	E	a	E	E	E	a	E	a	a	a	E	E	a	X	E	4

TABLE 11 (continued)

		Dellums	Hawkins	Burke	Young	Metcalfe	Collins[a]	Mitchell	Conyers	Diggs	Clay	Chisholm	Rangel	Stokes	Nix	Jordan	Index of Environmental Voting
13.	Mineral Strip	E	E	X	E	X	E	E	E	X	E	E	E	E	a	X	5.5
14.	Mass Transit	E	E	E	E	E	E	E	a	E	E	E	E	E	E	E	13.5
15.	Environmental Educ.	E	E	E	E	E	E	E	E	E	E	E	E	E	E	E	15
16.	Emergency Energy	E	a	E	E	a	E	E	E	E	a	E	a	a	a	E	6
17.	Delay Clean Air	E	E	a	E	a	E	X	E	a	a	E	E	a	E	X	3.5
18.	Energy Savings	E	E	a	E	E	E	E	E	E	E	E	E	E	E	E	13.5
19.	Energy Research	E	E	E	E	X	X	E	E	E	E	E	E	a	X	E	7.5

a. Collins entered House during session

b. Based on 14 votes only

c. For more descriptions of each vote, see Appendix B

E = "environmental" vote; X = "anti-environment" vote; a = absent, not voting cost of fuel (the environmental position on the Alaskan pipeline); in the costs of food (environmentalist opposition to the Eckhardt amendment to the toxic substances bill); in the price of autos (their opposition to the extension of air pollution deadlines). Higher prices for such essential goods, it could be argued, fall hardest on the poor, and therefore on many blacks. Nevertheless, even Representatives Dellums, Clay, and Mitchell, three of the more outspoken "nationalistic" blacks in Congress, voted against the Alaskan pipeline.

There are no indications that blacks actually blame environmentalists for the economic problems to which the poor, and particularly the racial minorities, are increasingly sensitive. While white public opinion spreads the blame for the energy shortages among the oil companies, the federal government, and the Nixon Presidency, blacks tend more often to single out the administration as primarily responsible (Gallup Opinion Index, 1974: 5; see also Wilson, 1974). It is in future relationships with black interests that environmentalism may well come into conflict, as environmentalists try to preserve the gains of the 1969-1973 legislation, in the face of demands for economic development and full employment. Energy shortages have higher issue salience for blacks than do environmental problems, and it should, logically, be harder in the future for black congressmen to follow their liberal ideological impulses into environmental voting. As explained previously, blacks have appeared to be "environmentalist" mostly by coincidence; their liberalism causes them to vote along with environmentalists on many bills, because of the high correlation between liberalism and environmentalism (Ritt and Ostheimer, 1974). Our present data show that black congressmen have been, by and large, in disagreement with black leadership outside of Congress–particularly the black intellectual community–at least on environmental issues. Perhaps this contradiction has been relatively unimportant up to now, given the low issue salience of environmental problems. Evaluating the voting performance of the Congressional Black Caucus, Robeck (1974: 11) concluded that black congressmen have, by and large, represented the black constituency effectively, or at least much more effectively than did white congressmen whose districts contained many blacks. But Robeck warned, as a proviso to his general conclusion:

> However, there is variation in the relationship in the two issue areas studied [civil liberties and social welfare], and probably in other issue domains as well. One suspects that as the issue becomes farther removed from the central concerns of blacks, that difference between black and white congressmen would become less predictable.

One might have suspected that, with the increasing issue salience of energy problems, black congressmen's "environmentalism" would decrease dramatically, at least for key economic issues: the "clean air" vote (Table 11, vote 17) led the authors to expect more such defections from environmentalism as the economic situation worsened. This would have been very logical as the middle-class issue of environment came into direct, short-run conflict with jobs and income for America's workers. Surprisingly, the consistently high levels of environmental voting found in past years among black congressmen did not diminish during 1974. In important bills dealing with land use planning (C.Q., 1974e), auto pollution controls (C.Q., 1974a), and strip mine regulation (C.Q., 1974f), the Black Caucus voted 100 percent "pro-environment." Indeed, all except Nix and Jordan (with Metcalfe, Conyers, and Clay absent), voted to abolish strip mining altogether (C.Q., 1974e). And with the execption of Nix (Diggs and Stokes absent), they voted with the environmentalist coalition in its attempt to delete $4.3 million from the A.E.C. appropriations bill for research on and evaluation of the use of underground nuclear explosives to stimulate the recovery of the shale oil and other minerals (C.Q., 1974b). Finally, on the Energy Supply and Coordination bill (C.Q., 1974b), which authorized the relaxation of several air pollution regulations, the Black Caucus was badly split. However, the environmentalist congressmen were also divided on this vote.

DISCUSSION

How can environmentalists develop positions that are more acceptable to blacks, and, possibly, gain the added support of one-tenth of the nation's voters? (The authors assume the gains of wooing the black constituency would outweigh the losses to the environmental movement caused by adding white backlash to the list of crosses they would have to bear.) The plight of America's cities represents a significant concern that is shared by the two groups. But it is supremely unfortunate for the course of cooperation between the two interests that their strategic solutions (even to those problems both interests specifically share) have diverged greatly. Environmentalists take a holistic view of their subject. According to the "no such thing as a free lunch" principle, all actions have environmental impact, and the deterioration of the biosphere must be tackled in a coordinated way. Applied to governmental philosophy, this has resulted in some degree in centralistic thinking, and in reliance on centralized bureaucratic structures and rules. However, during the 1960s, the thoughts of many black leaders were moving in the opposite direction: faith in federal government declined in favor of the assertion of local political control—black power over at least those levels of government that were within reach.

Given this discrepancy of approaches, and the failure of white environmentalists to recognize how sensitive blacks are to these issues of political power and representation, conflict was inevitable. A perceptive article by Edward Rodman (1970) clearly identified the central problem, depicting "environmental control [as] a new source of tensions between blacks and whites." Rodman agreed heartily that America's cities were environmentally unsound, and recognized that "black and brown people are the principal groups affected by these conditions." He stressed, however, that those wanting to help clean up the cities should recognize certain "basic realities," among which he listed "black consciousness" and "community control." Only the black community has the *will* to carry out the cleanup, which would necessitate deep social and economic reform, not simply cosmetic forms of environmentalism. Black control was the key:

> The creation of new environment regulatory agencies could become another obstacle to community development by injecting further red tape into an already hopeless process . . . the exclusion of blacks from panels, agencies, and commissions to deal with this problem is the first sign of the continuation of an old and consistent pattern [Rodman, 1970: 63]

The February 1974 attack by New York State Senator Joseph Galiber on Governor Wilson's fuel allocation plans has provided contemporary evidence that this continues to be a major divisive issue. Galiber called for citizen representation rather than unilateral, central control in fuel allocation decision-making (Cummings, 1974).

Environmentalists might be wise to recognize the direction indicated by these trends. Blacks are increasingly interested in controlling the decisions that affect the communities in which they live. But if and when blacks achieve power over their local decision-making machinery, they are, the data indicates, far more likely to treat health care, jobs, and schooling (to name a few crucial issues) as top priority items *before* they will turn to clean air, good water, and other "quality of life" concerns. Thus, if they wish to communicate with the black community, environmentalists must temper their desire for "logical" (that is, centrally coordinated) policies, and, instead,

(1) stress how serious environmental concerns are to the urban poor, even though air pollution may not kill people as fast as armed robberies do, and

(2) how local, black-controlled political machinery and effort can help to ameliorate the situation.

Finally, a deeper question arises from all of this. As Swan (1972: 70) puts it, "it is doubtful that air pollution will *ever* become an extremely high priority item with the very poor" (emphasis added). Only limited success may be possible in efforts to convince urban blacks that pollution is worth fighting. Environmentalists may wish to consider the advantages of coalition-forming: exchanging their support on black political issues for black support on environmental questions. This was in fact the strategy of community organizing efforts in Gary, initiated by the late Saul Alinsky. Such an approach will require abandoning the tendency to assume that blacks "ought to be concerned" about the urban physical environment, in favor of admitting that other issues are currently more salient to blacks. As compensation for the psychic costs of such a frank admission, environmentalists might ask themselves "What have we to lose?" Environmentalism is rather well identified by now as a "liberal" issue. The economic development constituency has already been lost, but another, very large, constituency stands to be gained.

Practical political considerations must be made if such a suggestion is to be taken seriously. Willard Johnson's analysis provides insights here. To Johnson, the real issues that concern blacks in their "environment" are poverty and humiliation amid a plentiful society. The only genuine solution is redistribution, and Johnson describes a variety of negative income tax that would benefit the 70 percent of the American people at middle or lower income levels. This is certainly a viable political base for any policy, but as Johnson explains, in spite of public acceptance of less obvious methods of public subsidies for the wealthy, the open and visible redistribution of income to "non-working" people goes against the grain of American values. The individualism of American culture—coupled with the racism of the majority, who would benefit economically, but would find the prospect of participating in the same system that was subsidizing the blacks too obnoxious to bear—would limit the actual support such redistributive policies might earn (1973: 178-185). Thus, an example of the supporting strategies environmentalists might consider is the income-distribution effects of taxation policy. Even more important they might give more attention to ways to reshape the values that have given America such a work complex. Of course, these may be unpalatable options for many environmentalists, many whom come to their interests through the hard sciences and are oriented to neither philosophy nor social policy. But, such political initiative might go far to impress upon blacks and other relatively disadvantaged groups that environmentalism *is* concerned with genuine human problems, not only with wilderness and endangered species.

Another question facing environmentalists is whether the emphasis on central coordination in decision-making has been overemphasized. This line of reasoning leads to the following question: Can progress in cleaning up the physical environment be made at all in communities where local versus distant control is a key issue—unless there is a large measure of public acceptance within the community concerned of the seriousiness of the threat that deterioration of physical environment represents. Public commitment and participation in environmental cleanup and control activities requires the presentation of more clearcut and dramatic evidence of the importance of poor physical environment as a casual agent for anti-social activity. Much technical research along these lines has already shown that human physiology is adversely affected by urban noise, air pollution, physical crowding, and aesthetically sterile surroundings. Environmentalists need to emphasize the human impact of these conditions particularly upon the urban poor, black and white, in America's cities. They must also offer specific recommendations as to the possible solution of these problems and counsel with black community leadership. Environmentalists might consider the possibility of working on a more applied level, possibly in conjunction with community health clinics or at least locating and publicizing them.

In view of the evidence presented in this paper, the gaps between environmentalism and black interests are, and have been, wider than are generally recognized at least among environmentalists. The mid-1970s are proving a frustrating time for most Americans economically. With the failure of the economic position of the black community to improve, the gulf will not be easy to bridge.

NOTES

1. The suspicion many American blacks hold for the population control sentiments expressed by environmentalists is also expressed by certain radical ideologists (Progressive Labor Party, 1971); it is paralleled on the international level by the leaders of some non-white countries who feel that population control rhetoric is directed primarily at their societies (Almeida, 1972: 43-44).

2. A cautionary note is in order at the beginning of this analysis of national sample survey data. National surveys usually include fewer than 200 black respondents from the 1,500 total sample. See the warning of the Harris organization (1969: 25).

3. The Congressional Black Caucus has been discussed in Poinsett (1973c) and Robeck (1974).

4. Black congressmen continued to be "unevironmental" on this issue when it came up again during 1974, with eight voting to build the Maine wilderness dam, five voting environmentally, and two absent (C.Q., 1974c; C.Q., 1974d).

5. It is possible that scoring absences as we did lowered black congressmen's voting scores not because they were anti-conservation, but because they did not care one way or the other about such issues. The effect is still the same.

6. It is ironic that data on the impact of the energy shortage layoffs on blacks may be increasingly difficult to collect as federal forms that do not denote workers' race or sex become prevalent.

REFERENCES

ALMEIDA, M.A.O. de (1972) "The confrontation between problems of development and environment." Internatl. Conciliation 586 (January): 37-56.

BAKER, H. A., Jr. (1971) "The environment as enemy in a black autobiography: Manchild in the Promised Land." Phylon 32, 1 (Spring): 53-59.

Black World (1972) "The national black political agenda: the Gary conference report." Vol. 21 (October): 27-31.

BRODY, E. B. (1974) "Mental health and population control." Mental Hygiene 58, 1 (Winter): 12-14.

C.Q. [Congressional Quarterly Weekly Report] (1974a) May 1 vote on Wyman Amend. Rejected 169-221. Vol. 32, 18 (May 4): 1138.

--- (1974b) April 30 vote on Roncalio Amend. to delete A.E.C. funds, rejected 190-207, and May 1 vote on Wyman Amend. to suspend air pollution controls, rejected 169-221. Vol. 32, 18 (May 4): 1136, 1138.

--- (1974c) June 6, vote on Conte Amend. Rejected 186-201. Vol. 32, 23 (June 8): 1526.

--- (1974d) "Dickey-Lincoln Project." Vol. 32, 24 (June 15): 1578.

--- (1974e) June 11 vote on federal grants to states for comprehensive land use planning. Rejected 204-211. Vol. 32, 24 (June 15): 1586, 1569.

--- (1974f) July 18 votes on Hosmer Amend. for regulation and reclamation, rejected 156-255, and on Hechler Amend. to end strip mining altogether within 54 months. Rejected 69-336. Vol. 32, 29 (July 20): 1914, 1917-1918.

CUMMINGS, J. (1974) "Crisis said to hurt blacks the most." New York Times (February 22).

DAVIES, J. C., III (1970) The Politics of Pollution. New York: Pegasus.

--- (1974) "Three mayors speak of their cities." (February): 35f.

--- (1973) "Ten best cities for blacks: survey suggests that blacks fare best in large urban areas." (November): 152f.

--- (1970) "Black mayors discuss state of the nation." (February): 76-84.

FARRELL, W. E. (1974) "Blacks fear layoffs and rising costs in energy crisis." New York Times (January 26).

Gallup Opinion Index (1974a) "Energy controls." Report 103 (January): 10.

--- (1974b) "Large families falling from favor here as well as abroad." Report 107 (May): 25-29.

--- (1973) "Energy crisis: speed limit reduction." Report 99 (September): 37.

GREGORY, D. (1971) "My answer to genocide." Ebony (October): 66-72.

GULLATTEE, A. M. (1973) "The politics of eugenics," pp. 82-93 in J. Robitscher [ed.] Eugenic Sterilization. Springfield: Charles C. Thomas.

HALLOW, R. Z. (1969) "The blacks cry genocide." Nation (April 28): 535-537.

HARE, N. (1970) "Black ecology." Black Scholar 1 (April): 308.

HARRIS, L. and Associates, Inc. (1969) "A study of the attitudes of the American public toward improvement of the natural environment." Study 1939A (July): 5.

HUDGINS, J. (1972) "Is birth control genocide?" Black Scholar 4 (November 12): 34-37.

JOHNSON, W. R. (1973) "Should the poor buy no growth?" Daedalus 102, 4 (Fall): 165-190.

JORDAN, V. E., Jr. (1974) "The energy crisis: for blacks, a disproportionate burden." New York Times (February 9).

KREGER, J. (1973) "Ecology and the black student opinion." J. of Environmental Education 4 (Spring): 30-34.

League of Conservation Voters (1961-1970, 1971, 1972, 1973) "How your congressman voted on critical environmental issues." (Washington, D.C.).

LOVERIDGE, R. O. (1974) "Air pollution and the political agenda: the issue, time, and policy choices." Presented at Western Polit. Sci. Assn. Annual Meeting, April 4-6, Denver.

MASON, B. J. (1973) "New threat to blacks: brain surgery to control behavior." Ebony (February): 62-72.

National Opinion Research Center (1973) "Spring 1973 general social survey." (July).

--- (1972) "Spring 1972 general social survey." (July).

New York Times (1974) "Report to black parley scores avoidance of minorities' needs." (March 16).

POINSETT, A. (1973a) "The politics of benign neglect." Ebony 28 (January): 33-40.

--- (1973b) "The joint D.C. center for political studies backs up black elected officials." Ebony 28 (April): 124-145.

--- (1973c) "The black caucus: five years later." Ebony 28 (June): 64-73.

Progressive Labor Party (1971) "ZPG: a fascist movement!" pp. 68-76 in D. Callahan [ed.] The American Population Debate. New York: Doubleday.

RITT, L. G. and J. M. OSTHEIMER (1974) "Congressional voting and ecological issues." Environmental Affairs 3, 3 (July): 459-472.

ROBECK, B. W. (1974) "The congressional black caucus and black representation." Presented at Amer. Polit. Sci. Assn. Annual Meeting, August 29-September 2, Chicago.

RODMAN, E. W. (1970) "The promise and threat of ecology: a black perspective." Yale Alumni Mag. (May): 61-63.

ROSENBAUM, W. A. (1973) The Politics of Environmental Concern. New York: Praeger.

SALAAM, K. (1974) "Our number one social critic." Black Collegian 4, 5 (May, June): 18-21.

SEWARD, J. N. (1974) "America's economic crisis: a black perspective." Black Scholar 5, 5 (February): 2-12.

SLATER, J. (1973) "Sterilization: newest threat to the poor." Ebony 28 (October): 150-156.

STRICKLAND, W. (1972) "The Gary convention and the crisis of American politics." Black World 21 (October): 18-26.

SWAN, J. A. (1972) "Public response to air pollution," pp. 66-74 in J. F. Wohlwill and D. H. Carson [eds.] Environment and the Social Sciences: Perspectives and Applications. Washington: Amer. Psychological Assn.

Time Magazine (1970) "The rise of anti-ecology." (August 3).

TOGNACCI, L. N., R. H. WEIGEL, M. F. WIDEEN and D.T.A. VERNON (1972) "Environmental quality: how universal is public concern?" Environment and Behavior 4 (March): 73-86.

TROP, C. and L. L. ROOS, Jr. (1971) "Public opinion and the environment," pp. 52-63 in L. L. Roos, Jr. [ed.] The Politics of Ecosuicide. New York: Holt, Rinehart & Winston.

U.S. Bureau of the Census (1974) Current Population Reports: "Fertility histories and birth expectations of American women." (June). Washington: U.S. Gov. Print. Office.

WALTERS, R. L. (1974) "Population control and the black community." Black Scholar 5, 8 (May): 45-51, and 5, 9 (June): 25-31.

--- (1972) "The new black political culture." Black World 21 (October): 4-17.

WATTS, W. and L. A. FREE (1973) State of the Nation. New York: Universe.

WILSON, E. J., III (1974) "The great energy gap, circa 1970-1990." Black Scholar 5 (March): 19-31.

APPENDIX A

DESCRIPTION OF LEGISLATION IN TABLE 9

1971 LCV Rating

1. *Family Planning:* Bill authorizing $180 million for family planning services. (HR 19318, November 16, 1970) E = "yes"
2. *SST:* Yates Amendment, D.O.T. Appropriations Bill, to delete $134 million for SST prototype development. (HJ Res 468, March 18, 1971) E = "yes"
3. *SST:* Boland Amendment to continue building SST. (HR 8190, May 12, 1971) E = "no"
4. *Mass Transit:* Giamo Amendment to restore funding for District of Columbia rapid transit system. (HR 8190, May 11, 1971) E = "yes"
5. *Mass Transit:* Giamo Amendment to restore District of Columbia metro funds when threatened a second time. (HR 11932, December 2, 1971) E = "yes"
6. *Channelization:* Reuss Amendment to prohibit funds for new stream channelization projects. (HR 9270, June 23, 1971) E = "yes"
7. *Radioactive Wastes:* Skubitz Amendment to delete $315 million that would liable A.E.C. to bury wastes in Kansas. (HR 9388, July 29, 1971) E = "yes"
8. *Amchitak Blast:* Mink Amendment to prohibit funds for Project Cannikan blast. (HR 10090, July 29, 1971) E = "yes"
9. *Dickey-Lincoln Dam:* Clark Amendment to delete funds for restudy of Dickey-Lincoln dam in Maine. (HR 19909, July 29, 1972) E = "yes"
10. *Alaska Wilderness:* Udall-Saylor Amendment authorizing 100 million acres for parks and wildlife. Alaskan Federation of Natives opposed the amindment. (HR 10367, October 20, 1971) E = "yes"
11. *Pesticides:* Dow Amendment to strengthen Pesticides Control Act. (HR 10729, November 9, 1971) E = "yes"
12. *Pesticides:* Dow Amendment to give any citizen legal standing to challenge E.P.A. regulation of pesticides. (HR 10729, November 9, 1971) E =yes"
13. *Pesticides:* Eckhardt Amendment to ban compensation to pesticide manufacturer for sales losses on banned products, if the company knew or could have foreseen danger of product. (HR 10729, November 9, 1971) E = "yes"
14. *Campaign Practices:* Pickle Amendment to weaken bill that would require radio and TV stations to sell time to political candidates at lowest unit rate, thereby helping less wealthy candidates. (HR 11060 and HR 11231, November 29, 1971) E = "no"
15. *Campaign Practices:* Danielson Amendment to eliminate requirement that candidates file financial reports with officials in their own districts as well as in Washington. (HR 11060, November 30, 1971) E = "no"

1972 LCV Rating

1. *Water Pollution:* Reuss Amendment to strengthen F.W.P.C.A. by setting 1981 deadline for industries to use "best available technology," and 1985 deadline for "zero discharge." (HR 11896, March 28, 1972) E = "yes"
2. *Water Pollution:* Abzug Amendment to delete from F.W.P.C.A. provisions weakening N.E.P.A. (HR 11896, March 28, 1972) E = "yes"

3. *Water Pollution:* Reuss Amendment allowing E.P.A. to veto state government permits to individual industries. (HR 11896, March 28, 1972) E = "yes"
4. *Water Pollution:* Mahon Amendment to require yearly congressional approval rather than E.P.A. contract authority for funding of waste water treatment plants. (HR 11896, March 29, 1972) E = "no"
5. *Water Pollution:* Ford Amendment to require public hearings to investigate industry war claims that pollution control costs require layoffs. (HR 11896, March 28, 1972) E = "yes"
6. *Recycle Sewage:* Vander-Jagt Amendment to direct E.P.A. through its water treatment grants to encourage recipient governments to use waste recycling. (HR 11896, March 29, 1972) E = "yes"
7. *Government Liability:* Bill allowing food industry to sue federal government for losses incurred from 1969 ban on cyclamates. (HR 13366, July 24, 1972) E = "no"
8. *Workplace Environment:* Dennis Amendment to delete $20 million from Occupational Safety and Health Act. (HR 15417, June 15, 1972) E = "no"
9. *Workplace Environment:* Findley Amendment to exempt very small firms from provisions of Occupational Safety and Health Act. (HR 15417, June 15, 1972) E = "no"
10. *Highway Trust:* Consideration of Anderson Amendment to free $700 million from Highway Trust Fund for use in urban transit systems. (HR 16656, October 3, 1972) E = "no"
11. *Three Sisters Bridge:* Abzug Amendment to delete provisions forcing construction of bridge through District of Columbia parks and historical sites. (HR 16656, October 5, 1972) E = "yes"
12. *Strip Mine:* Passage of bill imposing strict regulations on strip mining. (HR 6482, October 11, 1972) E = "yes"
13. *Water Pollution:* Override of President Nixon's veto of "expensive" F.W.P.C.A. appropriations. (S 2770, October 18, 1972) E = "yes"

APPENDIX B

DESCRIPTION OF LEGISLATION IN TABLE 11

1. *Mass Transit:* Anderson Amendment to allow urban areas to use $740 million from Highway Trust for mass transit. (S 502, April 19, 1973) E = "yes"
2. *U.N.E.P.:* Fraser recount on Wylie Amendment to reduce U.S. funds for U.N.E.P. (HR 6768, May 15, 1973) E = "no"
3. *Population Growth:* Hogan Amendment to forbid Legal Services Corp. to help women obtain legal abortions. (HR 7824, June 21, 1973) E = "no"
4. *Atomic Power Plants:* Podell Amendment to allow states to establish stronger regulations over atomic power plants. (HR 8662, June 25, 1973) E = "yes"
5. *Project Plowshare:* Roncalio Amendment to delete $3.8 million for Project Plowshare. (HR 8947, June 28, 1973) E = "yes"
6. *Geothermal Energy:* Lujan Amendment to fund geothermal energy research (HR 8947, June 28, 1973) E = "yes"
7. *Pesticides:* Bergland Amendment to restore the Department of Labor's authority to regulate the application of pesticides to protect farm workers. (HR 8860, July 16, 1973) E = "yes"
8. *Toxic Substances:* Young Amendment to prevent E.P.A. from implementing Act unless industry has the opportunity to cross-examine the agency on every ruling. (HR 5356, July 23, 1973) E = "no"
9. *Toxic Substances:* Eckhardt Amendment for better executive branch enforcement over toxic substances problems. (HR 5356, July 23, 1973) E = "yes"
10. *Alaska Pipeline:* Dingell Amendment to protect national parks, refuges, and wilderness areas from pipline construction. (HR 9130, August 2, 1973) E = "yes"
11. *Alaska Pipeline:* Dellenback-Owens Amendment to delete language saying that the bill complied with the National Environmental Policy Act. (HR 9130, August 2, 1973) E = "yes"
12. *Alaska Pipeline:* Eckhardt Amendment to restrict the N.E.P.A. exemption to the construction of the pipeline only, rather than to extend it to operation and maintenance. (HR 9130, August 2, 1973) E = "yes"
13. *Mineral Strip:* Melcher motion to suspend rules and pass bill to reimburse 32 ranchers whose grazing permits were cancelled on lands given back to the San Carlos Apache Indians. (HR 7730, September 18, 1973) E = "no"
14. *Mass Transit:* Wylie Amendment to protect Highway Trust Fund from "invasion" for mass transit development. (HR 6452, October 3, 1973) E = "no"
15. *Environmental Education:* Symms Amendment to shorten the life of the Environmental Education Program passed in 1970. (HR 3927, October 24, 1973) E = "no"
16. *Emergency Energy:* Hasting Amendment to suspend the Clean Air Act standards for auto emissions. (HR 11450, December 13, 1973) E = "no"
17. *Delay Clean Air:* Wyman Amendment to delay compliance with 1970 Clean Air Act standards until 1977. (HR 11450, December 14, 1973) E = "no"
18. *Energy Savings:* Bill authorizing G.S.A. to take energy-saving measures in Government buildings. (HR 11714, December 18, 1973) E = "yes"
19. *Energy Research:* Bill to create an Energy Research and Development Administration to study all methods of producing and conserving energy. (HR 11510, December 19, 1973) E = "yes"

JOHN M. OSTHEIMER teaches political science at Northern Arizona University. He earned his undergraduate and doctoral degrees at Yale University. Author of Nigerian Politics *(1973), he has written articles on African development and American environmental politics.*

LEONARD G. RITT also teaches political science at Northern Arizona University. He earned his B.A. and M.A. degrees from the University of Chicago and his Ph.D. from the University of Tennessee. He has published articles on American political attitudes, congressional voting behavior, and environmental politics.